AF228935

GET INVOLVED IN A ROBOTICS CLUB!

BY JENNIFER MASON

CAPSTONE PRESS
a capstone imprint

Published by Capstone Press, an imprint of Capstone
1710 Roe Crest Drive, North Mankato, Minnesota 56003
capstonepub.com

Library of Congress Cataloging-in-Publication Data is available on the Library of Congress website.

ISBN: 9781663958815 (hardcover)
ISBN: 9781666320305 (ebook PDF)

Summary: Wild about robotics? If so, a robotics club might be the right fit for you! Find out what it takes to join a robotics club or start your own, including information on membership, meetings, and activities. Together, you and your fellow members can participate, create, and, most importantly, have fun. Take the plunge, join the club, and get involved!

Editorial Credits
Editor: Ericka Smith; Designer: Sarah Bennett; Media Researcher: Svetlana Zhurkin; Production Specialist: Katy LaVigne

Image Credits
Alamy: Gabbro, 26; Associated Press: The Tribune-Democrat/Todd Berkey, 21; Courtesy of Anna Du, 15; Courtesy of Ronak Suchindra's Family, 5; Getty Images: M_a_y_a, 28, Peter Cade, 29, skynesher, 27, Thomas Barwick, 12, Tim Platt, 10; Newscom: ZUMA Press/Bruce R. Bennett, 20; Shutterstock: AlesiaKan, cover (top right), Anna Martynova, 19, Artgraphixel (blackboard), cover background and throughout, BAZA Production, 16, BGStock72, 23, Daisy Daisy, 6, Dmytro Zinkevych, 8, Gorodenkoff, 24, Karlis Dambrans, 22, Leika production, 14, Master Video, 4, memedozaslan photography, 11, mijatmijatovic, 13 (lined paper), MKE design, 13 (inset), Monkey Business Images, cover (middle), Oleg Erin, 9 (robot), Shmizla, 9 (paper), stas11, 25, Stokkete, 17, tdee photo cm, 7, UfaBizPhoto, cover (bottom), Vectorium (doodles), cover background and throughout, Veja, cover (top)

TABLE OF CONTENTS

Words in **bold** are in the glossary.

LOVE ROBOTS? JOIN THE CLUB

Robots are changing human life. They tackle chores. They assist with surgeries. They even help rescue people in a disaster. And that is just the start of what they can do!

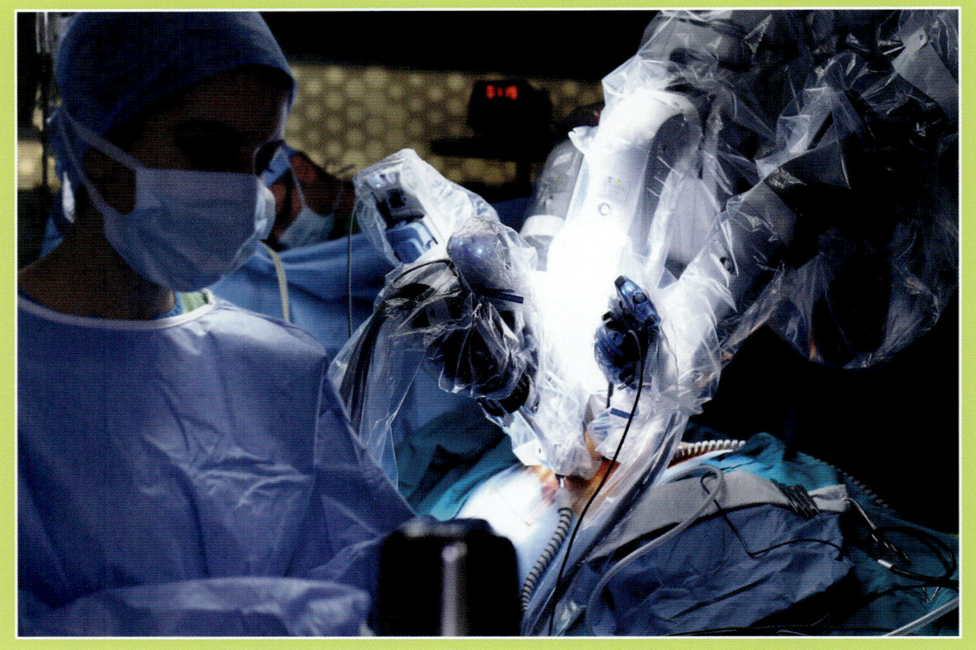

Building robots requires a large team with many skills. But you don't need access to a **laboratory** to get involved in robotics. A robotics club is the perfect place to start!

Joining a robotics club is a great way to meet other fans of robots. And you can learn new skills while building robots. Some clubs even enter competitions. Best of all, robotics clubs are fun!

No clubs to join in your town? No problem! Start your own!

Kids in Robotics

At age 13, Ronak Suchindra created his own online classroom, Kids Connect, because he was bored during the COVID-19 pandemic. Using Kids Connect, Ronak taught kids how to write **code** for robots. Ronak's class, and classes led by other kids on Kids Connect, were very popular. By 2021, about 2,500 kids from around the world had enjoyed 6,000 hours of activities. Ronak was also one of the top 20 finalists for *Time* magazine's Kid of the Year award in 2020.

CHAPTER 1

GETTING STARTED

Starting your own club isn't hard. You just need a passion for robots!

First, decide what you want to do in your robotics club. Do you want to practice building and coding robots? Do you want to enter competitions? Do you want to learn from robotics experts? You could focus on just one thing. Or you could do them all!

Once you've settled on a purpose, you can tackle the next steps. You'll need to find a **sponsor** and get kids to join your club. You'll also need to plan your meetings and raise money for your club.

Finding a Sponsor

You'll need an adult to sponsor your club. This person can help you find resources, build robots, and enter competitions. Ask a teacher, caregiver, or local business owner to sponsor the club.

Finding Club Members

Robots come in many different shapes and sizes. They can also do many jobs. Building them requires people with different skills and ideas. So keep club membership as open as you can.

Think about who should join your club. If you start a club at school, will students from other schools be allowed to join? If you want to enter competitions, are there age restrictions?

Once you decide who should join your club, find a way to share information about your club. Talk to kids in your school or neighborhood. Put up flyers in your community. Make sure you communicate what you want to do in your club and when and where you will have your first meeting.

Club Meetings

During your first meeting, the group can decide where to meet regularly. Garages are great for building robots. Local libraries or schools may have a free space. **Makerspaces** may have a room and tools.

The group should also decide when you will meet, how often you will meet, and how long your meetings will last. Will you meet after school? Will you gather weekly or monthly? How much time do you need for the activities you want to do?

Try It! Make a Flyer

Advertise your new club to the community with a flyer. Use this guide to get started.

First, insert your club's information where instructed. Make sure important words are the biggest. Use bold colors to grab people's attention.

Share digital copies of the flyer online. You can also print copies. Get help from an adult to put up your flyers at school and around town.

Funding Your Club

Clubs have costs. For robotics clubs, costs might include supplies and entry fees for competitions. How will you pay for these costs? Fundraising—selling things or asking for donations—is one way to raise money. Having members pay club **dues** is another option. Once you've figured out what you want to do in your club, talk with club members and your sponsor about how to cover the costs.

Bots, Meet Your Makers

Club meetings are where the fun happens. Members can rave about robots, learn from experts, munch snacks, and of course . . . make robots!

A successful first meeting is an important step. Here are a few helpful tips:

- Break the ice with a game. This will help people feel less shy.

- Pick a club name. Aim for a name that's just a few words. Express your club's purpose and spirit.

- Pass around a snack sign-up sheet. Building robots demands fuel!

- Decide on club **roles**—who will help lead the club.

Divide and Conquer

Big projects are easier when they are divided into smaller tasks. In a robotics club, members can take on different roles and share duties.

Clubs often have four roles for leaders:

1. A **president** directs the club and runs meetings.

2. A **vice president** steps in when the president is away.

3. A **treasurer** collects dues and pays for things. (Your sponsor should help with these tasks.)

4. A **secretary** sends meeting reminders, tracks new members, and advertises the club.

You can come up with more creative titles, like Bot Commander in Chief, Vice Overlord, Coin Collector, and Keeper of the Scrolls.

Members are elected for each role. And they serve for a set amount of time. That way, others get a chance to lead the group.

Meeting Agendas

An **agenda** is an important tool for keeping meetings organized. It helps you make sure everything gets done by the end of your meeting.

Try It! Make a Meeting Agenda

Use this example to create agendas for your club meetings.

CHAPTER 2

BUILD AND BATTLE BOTS

So, you started a robotics club. Nice work! Now it's time to build some robots! If you have raised enough money, buying robotics kits is a great way to start learning about bot building. They have all the parts you need to construct a robot. They may also include tools.

If kits are not an option, you can build robots from repurposed materials. Even professional robotics companies reuse everyday items.

A HOMEMADE SEA BOT

As a sixth grader, Anna Du built a remotely operated vehicle (ROV) that roamed under crashing tidal waves. She used PVC plumbing pipes and foam swim tubes. The ROV used special light signals to track down toxic plastics in the sea. In 2021, she used artificial intelligence to predict places where microplastics are located in the ocean.

Hunting for Supplies

Many of the parts you'll need to build a robot might be in your home right now! With a battery and a mini-motor, just about anything transforms into an awesome robot. For example, a motor driving a toothbrush makes a zipping, zooming bug-bot.

Hunt for junk in your home. Then gather tools and materials like screwdrivers, pliers, scissors, zip ties, and a hot glue gun. (You'll need help from an adult to use some tools safely.)

Ask local businesses to **donate** costly supplies like the laptops you'll need for programming. A school or library may have a computer lab you can use. If you need 3D printers to make robot parts, check with makerspaces.

Try It!   Plan a Club Scavenger Hunt

One of your club's first fun activities can be a scavenger hunt! As a group, search one another's homes or the neighborhood for useful scrap materials. Look for old toys or gadgets with motors, gears, tubes, pumps, or springs. Keep your eyes peeled for things that spin, light up, beep, or talk. Cardboard and cartons are perfect for robot bodies. Jar lids and old CDs make great wheels.

Join a Competition

Robotics competitions are contests between different robots. People form teams to build the robots that will compete. In most cases, all the robots face the same challenges. But which robot best gets the job done in time? That's a lot of pressure.

Performing well under pressure is a useful skill. And it is one your club members can master when you compete together. Rather than designing and building at your own pace, contests drive teams to meet a **deadline**. Deadlines add pressure. But they also spark creativity!

To win a competition, the best teams communicate and cooperate. These skills also create stronger friendships and stronger clubs.

Competitions also push teams to learn new technical skills. These skills might include working with wobbly tilt switches or self-driving programs that use a global positioning system (GPS).

Game Changers

There are many competitions, big and small, for your club to choose from. The For Inspiration and Recognition of Science and Technology (FIRST) competitions are popular international contests.

Each year, the FIRST competitions involve new challenges. In 2020, teams had to create robots that were able to defend an imaginary planet from asteroids. Robots had to move across a furry planet surface. They also had to grab foam balls and toss them through target hoops faster than a rival team.

Other popular competitions include the VEX Robotics competitions, the International Robot Olympiad, and the RoboNation challenges. Some contests seek the robot that can finish an obstacle course the fastest. Others expect robots to perform tricky tasks.

For many clubs, competitions are game-changing experiences. Winners sometimes receive prize money or scholarships—money for college. Check with your school, library, nearby colleges, or makerspaces for local contests.

CHAPTER 3

REACH OUT AND GROW

Members might need to learn the basics before building robots. Or they might just want to learn new skills throughout the year. Your club can use many resources to learn about robots. Libraries offer helpful books. The internet is packed with how-to videos. And many robotics experts enjoy sharing their knowledge with others.

Talk to Roomba Engineers

Ever heard of the Roomba robotic vacuum? In 2002, this cleaning robot changed how people interact with robots. The engineers who build and program Roombas love sharing their knowledge with students. They can visit in person or virtually.

Be Our Guest

Your club can learn a lot by listening to robotics experts. Invite local robotics experts to speak to your club. Start with professionals who live in your community. Robotics professors from local colleges are good options.

You can also **poll** club members to come up with a list of potential guest speakers. Your sponsor can help you find experts and their contact information.

Set a date and time for your guest speaker's presentation. Thirty minutes is a good target. It leaves time for follow-up questions.

As a club, create a list of questions you'd like to ask. Here are some ideas:

- Why did you decide to make robots? What were some steps you took to have this **career**? What advice do you have for people who want to become roboticists?

- What types of projects are you working on now? Why are those projects important?

- What are some of the challenges you've faced in building robots? How did you solve those problems?

Experts often enjoy sharing what they do. With an adult's help, start by sending a polite email to a potential guest speaker. Explain a little about your club. Let them know which topics you're interested in. Give them information about your meetings and how long you'd like them to speak. Find out if the guest can visit in person or virtually. Use the guide below to help you write an email to a robotics expert in your area.

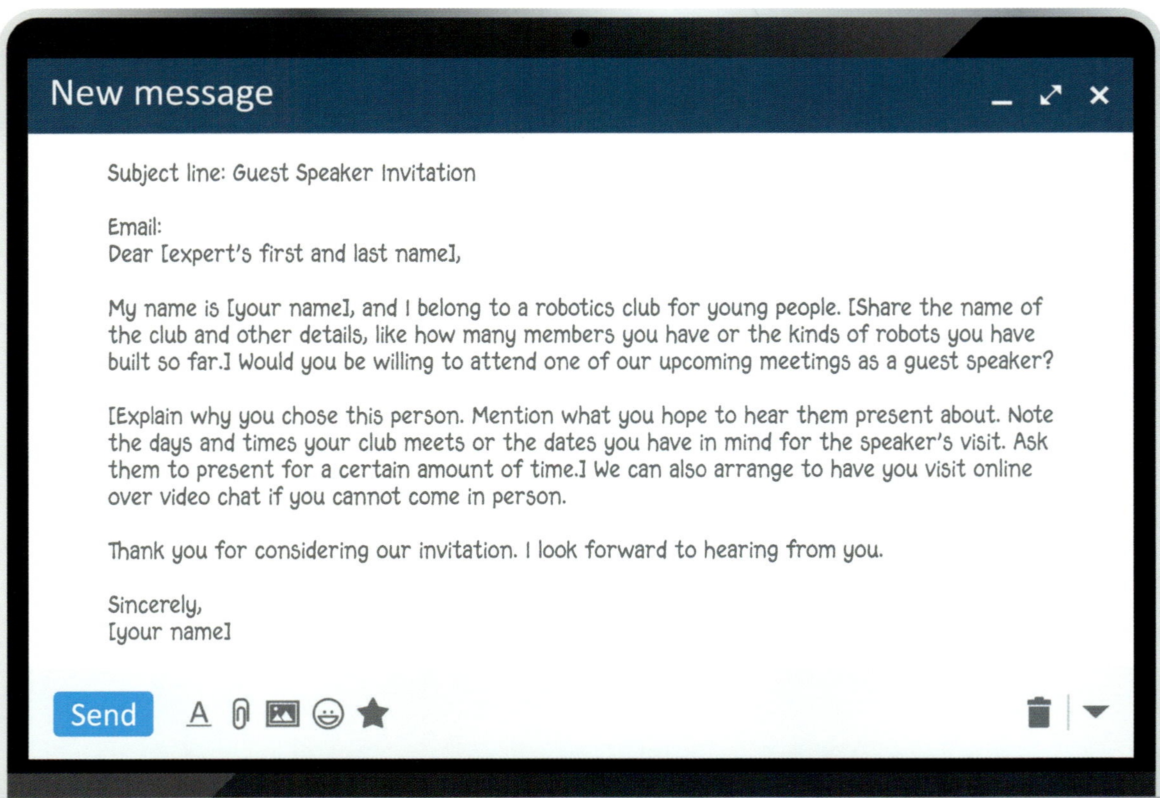

Community Events

Half the fun of robotics is showing off what you've built! Hosting a community event can be a great way to share your club's skills and passion with others. Sharing knowledge can also help you attract new club members.

Your event can be held in the same place as your club meetings. Or ask to use a large, open room at your school or library. These spaces are perfect for a community **showcase** to display your club's robots.

Plan an event as you would a club meeting. Use an agenda to organize activities. Advertise the event with a fun flyer. Set out plenty of snacks and drinks.

Demonstrate what the robots can do. Explain how they were built. And encourage guests to ask questions. You might even turn the showcase into a **festival** and invite robotics clubs from nearby towns!

Time to Celebrate

Hardworking clubs deserve a party! Take time to celebrate your achievements. Maybe you worked really hard for a robotics competition. Maybe you put together a great showcase. Whatever it is, take time to recognize that effort.

At the end of the year—or school year—host a party. You can also thank club members with participation awards. Think of fun names, like the "Code Champ" award. Awards can come with gifts too, like T-shirts or gift cards.

You can also plan something fun, like a sleepover. Taking time to just have fun will help your members build friendships and work together better.

Starting your own robotics club is an exciting challenge. You can make new friends, learn new skills, and get hands-on experience building cool bots!

GLOSSARY

agenda (uh-JEN-duh)—a list of things that will happen

career (kuh-REER)—the type of work a person does

code (KODE)—a system of words, letters, symbols, or numbers used instead of ordinary words to send messages or store information

deadline (DED-line)—the time by which something needs to be completed

donate (DOH-nayt)—to give something as a gift

dues (DOOZ)—a regular payment made in order to be a member of a club

festival (FES-tuh-vul)—an organized series of events

laboratory (LAB-ruh-tor-ee)—a place where scientists do experiments and tests

makerspace (MAY-ker-spayss)—a place where people go to design and build projects and share information

poll (POHL)—to ask people a series of questions in order to find out what a group thinks

role (ROHL)—a person's job or duties

showcase (SHOH-kays)—an event that displays the good qualities of something in an attractive or favorable way

sponsor (SPON-sur)—a person who is responsible for something

READ MORE

Katovich, Bob. *Awesome Robotics Projects for Kids: 20 Original STEAM Robots and Circuits to Design and Build.* Emeryville, CA: Rockridge Press, 2019.

Smibert, Angie. *Artificial Intelligence: Thinking Machines and Smart Robots with Science Activities for Kids.* White River Junction, VT: Nomad Press, 2018.

Sobey, Ed. *Robotics Engineering: Learn It, Try It!* North Mankato, MN: Capstone, 2018.

INTERNET SITES

Code.org
code.org

Makey Makey
makeymakey.com

Raspberry Pi
raspberrypi.org

INDEX

ABOUT THE AUTHOR

Jenny Mason is a story hunter. She explores exotic countries, canyon mazes, and burial crypts to gather the facts that make good stories. She is an active member in clubs for books, tennis, martial arts, history, and dancing.